# Rocketplanes to Space

by Patrick H. Stakem

(c) 2017, 2022

2nd Edition

Number 10 in the Space Series

# Table of Contents

Introduction..........4
Author..........4
A note on Units..........6
Early Efforts..........9
   Silbervogel..........11
   World-War-II Rocket powered fighters..........13
      Germany..........13
      Japan..........22
      Soviet Union..........23
      Early U. S. and Allied effort..........25
   Post war and Space Age..........26
U. S. Rocketplanes..........27
   X-1..........27
   X-24..........28
   X-30..........29
   Blackstar..........30
   HL-20..........31
   X-15..........31
   X-20 Dynasoar..........33
   X-33 and X-34..........35
   X-37..........35
   XP-79..........37
   STS-Shuttle..........37
   X-38 Crew Return Vehicle..........42
Other Nation's Spaceplane Projects..........43
   Buran – the Russian Shuttle..........43
   Russian BOR-4..........44
   British Hotol..........45
   Skylon..........45
   German Sanger II..........46
   Indian Avatar..........47
   French Hermes..........48
   ESA SOAR..........48
   Airbus Defense and Space Spaceplane..........48
Commercial Ventures..........49

- Virgin Galactic..................................................49
- Sierra Nevada's Dreamchaser................................50
- Scaled Composites..............................................51
- Rocketplane Kistler.............................................53
- XCOR ...............................................................54
- Orbital Sciences Prometheus.................................55
- Bristol Spaceplanes.............................................56

Wrap-up..................................................................57
Bibliography............................................................59
Resources................................................................68
Glossary of terms......................................................70
If you enjoyed this book, you might be interested in some of his others as well. ......................................................77

## Introduction

This book covers the topic of Spaceplanes, that are launched to orbit, and return to a runway landing. The most common example of a Spaceplane is the Space Shuttle. This discussion is limited to crewed vehicles, and is not comprehensive. This is book ten in a series on Space related topics. Predecessors to the Spaceplane was the Rocket aircraft, starting in Germany prior to World War-2. The technology and some of the experts were brought over to the United States, and the Bell X-1, the first aircraft to exceed the speed of sound, and the X-25, the first winged craft to reach Space, resulted. These efforts influenced the Space Shuttle. There are ongoing rocket-powered winged plane projects in several countries, and by several commercial companies as well.

## Author

Mr. Patrick H. Stakem has been fascinated by the space program since the Vanguard launches in 1957. He received a Bachelors degree in Electrical Engineering from Carnegie-Mellon University, and Masters Degrees in Physics and Computer Science from the Johns Hopkins University. At Carnegie, he worked with a group of undergraduate students to re-assemble and operate a surplus missile guidance computer, which was later donated to the Smithsonian. He was brought up in the mainframe era, and was taught to never trust a computer you could lift.

He began his career in Aerospace with Fairchild

Industries on the ATS-6 (Applications Technology Satellite-6), program, a communications relay satellite that developed much of the technology for the TDRSS (Tracking and Data Relay Satellite System). He followed the ATS-6 Program through its operational phase, and worked on other projects at NASA's Goddard Space Flight Center including the Hubble Space Telescope, the International Ultraviolet Explorer (IUE), the Solar Maximum Mission (SMM), some of the Landsat missions, and Shuttle. He was posted to NASA's Jet Propulsion Laboratory for MARS-Jupiter-Saturn (MJS-77), which later became the *Voyager* mission, which is still operating and returning data from outside the solar system at this writing. He initiated and lead the Flight Linux Project for NASA's Earth Sciences Technology Office. During a career as a NASA support contractor from 1971 to 2013, he worked at all of the NASA Centers.

He has the NASA Shuttle Program Manager's Commendation Award as well as the NASA Apollo-Soyuz Test Program Award, and two NASA Group Achievement Awards.

Mr. Stakem was affiliated with the Whiting School of Engineering of the Johns Hopkins University, the Graduate Computer Science Department at Loyola University in Maryland, and Capitol Institute of Technology. Mr. Stakem supported the Summer Engineering Bootcamp Projects at Goddard Space Fight Center for 2 years. He has developed and presented

Cubesat courses, and is active in addressing STEM.

## A note on Units

I am fairly conversant in both English and Metric units (what is the metric equivalent of furlongs per fortnight?). Metric (SI) is mandated for NASA usage now, for interchangeability with our partner space faring nations. When a lot of the legacy flights discussed here were flown, English units were the norm. I have tried to keep the units comparable to the mission at the time. Conversions are easy enough, but units conversion is a source of error. It's not what you know about units and measurement, its how you think. And, I still think English units (even the English use Metric now), and convert in my head or on my phone.

For scientific/engineering work, the Metric system is well thought out. For artisans, the English system served well, as most units were divided by 2, which is easy. Fold the cloth. Hopefully, when we are all taught Metric first, some one will still remember the conversions. You just need a slide rule....

# Definitions, and scope.

A Spaceplane is designed for winged flight in the atmosphere, and uses thrusters for maneuvering in space. Outside the atmosphere, the wings are just decorative. The obvious example of a space plane is the Space Shuttle.

Going up above the atmosphere, the wings are mostly in the way. There are several ways to get to orbit, all involving rocket engines. You can't do it completely with an air-breathing engine, because it will rapidly run out of air to breath. (And, I'm not discussing space elevators here. They work best if there is no atmosphere, like on the moon). The Shuttle had rocket engines, fed from the large external fuel tank that was jettisoned when empty. It also had two large solid rocket engines, which were, essentially, a first stage. There were also jettisoned when finished. The X-15 rocket plane flew to space, but only after it was taken to altitude under the wing of a mothership. It didn't have enough fuel to take off from the ground and achieve extreme altitudes. It's "first stage" was a B-52 aircraft. A similar technique was developed by famed designer Burt Rutan, for Scaled Composites. The purpose-built carrier vehicle, White Knight, carries the smaller SpaceShip to a high altitude. At this point the two vehicles separate, and the White Knight flys back to a runway landing as the smaller craft continues to orbit with its rocket engine. Rutan is distinguished by having five of his planes on display in the Smithsonian National Air and Space Museum, including Voyager, which in 1986 was the first airplane to fly around the world un-refueled, and SpaceShipOne, which in 2004 became the first private rocket plane ever to put a human into space.

The spaceplanes we are most interested in are Crewed vehicles. That has to do more with the return from orbit than getting there in the first place. Anything put into orbit requires a large expenditure of energy to do so. We need large rockets to put a payload into orbit. Once there, for all intents, it stays there. Let's ignore orbital decay for a while. When the payload wants to return to the surface, it has to get rid of a lot of energy that it accumulated on the way up. Consider driving a car up a steep mountain. From the top, you can drift down the other side for quite a while, without expending anything but your potential energy, which you have at the top of the hill. So, the returning craft is at the top of a few hundred miles of "hill." It has a lot of potential energy to get rid of. Coming down through the atmosphere at speed will cause the craft to heat up to the point of melting. The general solution was ablative heat shields on the early capsules, and ceramic tiles on the bottom of the Shuttle. Besides the potential energy of position, an object in orbit is moving at a high rate of speed, that has to be reduced to zero when it lands. With a capsule returning to land, as the Russians prefer, you can use retro-rockets and a parachute. When returning to an ocean landing (which the U.S. did), you use the chute, but don't need the retro-rockets. If your craft is winged, you will need the heat shielding on the bottom, but you can get rid of a lot of energy with maneuvers, just as an airplane does in landing. The spaceplane can use an ordinary runway, assuming its long enough.

We'll see that the early rocketplanes required one to be an accomplished glider pilot, since the landing was done unpowered. That means you have to get it right the first time. Rocket planes spurred the development of ejection seats, which would later be required for high performance jets. There is a weight penalty, but when you need one, they are handy.

## Early Efforts

This section discusses the early efforts in applying rocket-power to aircraft. Air-breathing engines run out of air before they reach their maximum possible altitude. Rocket propulsion, particularly liquid rocket engines, solve this problem. They carry, in addition to fuel, a supply of oxidizer. Liquid fueled engines are preferred to solid fueled, since the liquid engine can be shut down, restarted, and throttled.

I should mention here the concept of Mach number. Mach 1 is the speed of sound, at the ambient pressure and temperature. At mean sea level, in dry air at 15 degrees C , sound travels at 340.3 meters/second. Fluid Dynamics engineers use the Mach number to determine when the fluid flow becomes incompressible. Other than that, it is just a measure of speed. Subsonic refers to speeds less than Mach 1, and supersonic, to speeds above Mach 1. Trans-sonic is a narrow region around Mach 1, where the air flow is transitioning from sub-to-supersonic. Hypersonic refers to Mach 5-10. The speed of sound depends on the medium, its temperature, and pressure.

One big issue with the early efforts was that the rocket plane would go much faster than the propeller planes of the day. They were even capable of going faster than "the sound barrier." No one knew what would happen then, as there were no wind tunnels to test at that speed. Another big concern was aerodynamic heating. The sound barrier was the big unknown.

In doing research for this book, the author found thirty-four rocket plane projects from 1928 to the end of World War-2. The breakdown by country was, Germany-16; Italy-1; USSR-7; Japan-4; France-2; and the U.K.-4.

All of these reached the design phase. Some were drop tested with no engines; some flew with a jet engine, some had both a jet and a rocket engine. I am not going to discuss all of these, but just what I consider the significant ones, that lead to the later development of the X-15, and the Space Shuttle.

Possibly the first rocket powered powered aircraft was the Lippisch Ente (Duck, in German), circa 1928, in Germany. It was a sailplane with an added rocket engine, flow by Pilot Fritz Stamer. The Ente, like most of Lippisch's designs, was tailless, so fitting the rocket engine was easier. The rocket engines used were solid propellant, 44 pounds of thrust, ignited by a switch in the cockpit. The second engine was ignited by the pilot after the first burned out. On the second flight, they decided to use both at once. One of them exploded, and set the plane on fire. The pilot brought it down successfully, and abandoned ship hastily. The plane was never rebuilt.

The Opel Rak.1 was also a sailplane, fitted with rocket propulsion. Opel, the car manufacturer, experimented with rocket powered cars, as a publicity stunt, and advertising. Opel himself flew the plane almost a mile, but the craft was heavily damaged upon landing.

Silbervogel

Eugen Sanger, a doctoral candidate at Vienna Polytech in 1933, proposed a Mach-10, 160 km altitude capable glider. He published his Techniques of Rocket Flight in 1933. He refined his project to a hypersonic boost-glide vehicle in 1934. It was designed to achieve Mach 13 at engine cut-off. It was then capable of a 5,000 km glide at Mach 3.3 and 50 km. This is when every other plane was using one or more propellers.

He produced a paper later in World War-II, entitled Concerning Rocket Propulsion for Long-Range Bombers. This described a 28 meter long spaceplane with a 15m wingspan. The top speed was to be 21,800 kph, and a range of 23,400 km. It was to be launched by a rocket powered sled, using liquid oxygen and alcohol. The plane's rocket motor used liquid oxygen and kerosene.

Sanger sent a proposal to the German Ministry of Aviation in 1941, who promptly filed it. He went on to design and work on a ramjet interceptor/fighter plane, the Skoda-Kauba Sk P 14.01. This was to be powered by an engine he designed. The craft was to have been launched by booster rockets to a speed where the ram jet would operate. The top speed was calculated to be 1000 km/h. The first model was to have a wingspan of 7 meters, and a length of nearly 10. This plane was developed in 1945 to address the mass bombings of Germany during the war, but it never made it to service. The Germans were trying to find a way to counter the Allies' thousand plane raids. Late in the war, the British and Americans could field that many planes, daily, day and night.

The Silbervogel (German: Silver Bird) featured a lifting body architecture, where the entire lower surface of the craft acted as a wing. It was sub-orbital, and achieved long distances by skipping along the upper atmosphere. He also introduced the concept, in use today, of cooling the rocket nozzle and pressurizing the fuel, by circulating it around the nozzle in tubes.

The project was adopted by the Luftwaffe as the Amerika Bomber. It would launch from Germany, bomb New York, and fly to Japan for recovery. The scary part is, had Germany succeeded in constructing it, and in constructing an atomic bomb, history would have been very different. After the war, Dr. Walter Dornberger, the military officer in control of the V-1, V-2, and similar rocket projects came to the United States to work with the von Braun team at the Redstone arsenal in Huntsville, Alabama. He carefully referred to the Silbervogel as the Antipodal Bomber. The Russians worked on a rocket-powered sub-orbital bomber influenced heavily by Sanger's design, in the late 1940's – 1950. Luckily, Sanger had made an error in a heat flow calculation, that would have resulted in the craft being destroyed upon reentry. The early work by Sanger in Germany was applied to many of the subsequent German rocket projects.

Principles of the Silbervogel influenced the U. S. rocket planes such as the X-15 and the X-20. Sanger continued work on key aerospace concepts in Germany, designing a ramjet powered spaceplane in the 1960's. He also did work in laser propulsion and solar sailing. In 1985, Messerschmitt-Bolkow-Blohm used the Sanger concept in a study of a horizontally launched, two-stage-to-orbit plane.

Sanger can certainly be called the Father of the Space Plane, as he had worked out most the the details by 1933.

# World-War-II Rocket powered fighters

Germany

Rocket powered aircraft were flying in Germany in the 1930's, thanks to some good theoretical work, good engineering, and a loophole in the Armistice for World War-I. Germany was banned from developing any powered aircraft, which meant, at the time, anything with a propeller. To keep a cadre of pilots available, the sport of gliding became popular. They couldn't tow them to altitude with a plane, so they launched them downhill, or with a long tether that was wound in quickly. A rocket engine in a glider would enable the craft to get to altitude on its own. It bent the rules, but just a bit.

The ME-263 was a single seat rocket powered fighter, derived from the ME-163. Three were built, but none were ever flown, due to the end of the war. They were launched by towing to altitude by a larger plane. They could reach a maximum speed of 590 mph, and had a range and endurance of 78 miles, and 15 minutes.. they had twin 30 mm cannons.

Max Valier was working with car manufacturer Opel to get a rocket powered car going, as a publicity stunt for the car company. They started out using solid fueled rockets in 1928. First tests were disappointing, but they finally got to 125 mph. On railroad tracks, another model got to 180 mph, and used retro rockets to brake. Valier became a proponent of liquid fuel, and influenced a budding rocket pioneer, Wernher von Braun. Valier was killed when a rocket motor explosion drove a sharp splinter through his heart. Opel's company dropped the rocket car experiments.

The Heinkel 112 was the first aircraft to fly with a rocket engine, using von Braun's design. This was in the summer of 1937. It used liquid oxygen and alcohol, the same fuel that von Braun would use in his V-2 rocket some years later. The Heinkel P.1977 was a interceptor developed under the German Emergency Fighter Program, a response to heavy day and night bombing by the British and Americans.

Helmut Walter was also experimenting with liquid rocket engines. He used the more dangerous methanol/water/hydrazine fuel with hydrogen peroxide. One of his engines was tested in a Heinkel-111 aircraft as a take-off enhancement. Later, a HE-176 was towed behind a turbocharged 7.6 liter Mercedes car for take off, then it lit off the liquid engine. Evidently, the Allies from World War-I were not aware of this work, or didn't consider it important. The HE-176 was the first aircraft to fly with a liquid rocket engine. The earlier Espenlaub E-7 and Opel-Sander Rak-1 used solid fuel engines. Design of the HE-176 was begun in 1936. It used the Walter HWK-R1 engine. It was 6.2 meters in length, with a wing span of 5 meters. Three units were fitted with a von Braun liquid oxygen/alcohol engine. It flew in 1937. Work was officially stopped on the project in late 1939.

The German rocket planes were not going to space, but were designed as bombers and fighters that would be much faster than any of their propeller-driven adversaries. These efforts resulted in operational rocket planes, too little, too late. They did heavily influence later rocket craft, particularly in the United States. Operationally, against bomber squadrons with fighter escort, the plane's were of limited effectiveness, due to the short burn time of the engine. After the fuel was used up, the planes dove for the ground at high speed, and glided to a landing. This was noticed by the fighter-escorts, who managed to follow them down and blow them out of the sky as they prepared for landing.

Generally, the planes could not take off unassisted, so they were towed or carried by a larger plane. They could glide to a landing, after their mission was complete, or the fuel ran out..

The Heinkel HE-176 first flew in 1939, as the world's first rocket aircraft. Unfortunately, all of the documentation for the project were destroyed during the war, and only two pictures survive. There were actually deployed against Allied bomber formations late in the war, but proved almost as dangerous to the pilots as to their prey. Engine malfunctions and the dangerous nature of the fuel and oxidizer were a threat to the pilots and ground crew.

Warsitz, in his book about his father, Erich Warsitsz, the test pilot of rocket aircraft for the Luftwaffe, has his father mention that there were using data from a 1937 U.S. NACA Report, "Tests of 16 Related Airfoils at High Speeds."

The DFS-194 was a tailless design, originally with a pusher propeller. It was transitioned to a rocket plane, with first flight in 1940. It used a Walter R I-203 liquid fuel engine. It managed 340 mph on its first flight. It led to the Messerschmitt Me-163. That was the first aircraft to achieve 1,000 km/hr in level flight, not quite to the sound barrier. It was also the first mass-produced rocket plane.

The HE-162 Volksjager was a result of Germany's Emergency Fighter Program, built by Heinkel. It was constructed of wood. The ME-262A and the AR 234B were already in service, utilizing scarce strategic materials, and available rocket engines. It was introduced too late to make any impact on World War-II. It was originally powered by a turbojet engine, but a BMW 718 liquid fueled rocket engine was added, for speed bursts in combat. The fuel was diesel, and the oxidizer was red fuming nitric acid (one of the author's all-time favorites).

The ME-163 Komet was the first operational rocket-powered fighter aircraft. It was a tail-less design, dated back to 1929. It was also the first aircraft to exceed 1,000 km/hr, not quite the speed of sound. This record was kept secret. It was not pushed any faster, as the aerodynamics of the trans-sonic region was not well understood, and the necessarily wind tunnels for testing were not available. This speed record was not surpassed until late 1947, when an American jet went faster. In July 1944, a German pilot achieved 1,130 km/hr. This was vastly faster than the Allies' bomber fleet, and escort fighters. Over 300 of the planes were built, starting in 1941. Slave labor in concentration camps was used for their construction. The follow-on ME-163b was tested at the rocket range at Peenemunde. It was armored, and had 200 gallons of T-stoff, and 110 gallons of C-stoff in the wings. That gave 4-5 minutes of powered flight, after which the plane became a glider. There was no tail, but the craft used elevon control. Famed Avatrix Hanna Reitsch flew the craft, but crashed it. The original plane was equipped with 20mm cannon, upgraded to 30 mm in the B model, with 60 rounds of ammunition.

The original engine was from Walter, the R-1-203, using a mono-propellant, stabilized high-test hydrogen peroxide (above 85%). This fuel is extremely dangerous and unstable, and almost anything else acts as a catalyst. This engine was used operationally, but the propulsion unit was more of a danger to the pilot, than the plane was to the bomber fleets. Later, the engine was switched to a bipropellant using hydrazine and alcohol, with the oxygen from the decomposition of the hydrogen peroxide. This may have been more dangerous than the previous. The two liquids were hypergolic, meaning they reacted violently when mixed. No ignition source was needed. Ground handling of the liquids was possibly more dangerous than flying the plane. The pilots did not have pressure suits, which limited the ceiling of the aircraft. They did have an oxygen mask. In one terrible accident, the oxidizer, which reacts with organic matter, leaked into the cockpit at landing, and dissolved the pilot.

The plane had wooden wings, no tail, and only elevon control. The onboard tanks gave about 4-5 minutes of powered flight, and it landed as a glider. The engine developed around 3,750 pounds of thrust. The plane had set a world speed record of 572 miles per hour at the Peenemunde test facility, but this information was kept secret until after the war. In 1941, it reached 623 mph, Mach .84. Not much was known about aerodynamics beyond that point, as that exceeded the capability of contemporary wind tunnels. They suspected but did not know there was a "sound barrier." This is simply a change in the aerodynamics. Some aircraft controls are reversed in their effects. Drag and vibration increase abruptly. These effects are now well understood, and addressed by design.

The captured ME-163, circa 1941, influenced post-war aircraft research by NACA (predecessor to NASA). The designs of the F102A, the F106A, and the B-58 aircraft were directly influenced, as was the experimental X-1 and X-15.

The DFS-228 was a rocket-powered reconnaissance aircraft. By the end of World War-II, only two un-powered prototypes had been flown. It was essentially, a sailplane that could launch itself to high altitudes. It became a project of the German Aviation Ministry. It was to use the Walter HWK 109-509 engine, the same as used in the ME-163 and the Ba349. The prototype was captured by U.S. Troops in 1946, and was sent to the U.K. The Walter engine used C-Stoff and T-Stoff, which are hybergolic. The ratio was 3 parts T to 1 part C. The engine produced about 17 kilo-newtons of thrust (3,800 lf-f). A turbo-pump scheme was used to get fuel and oxydizer into the engine. This was the approach used in the V-2 rockets, and most launch vehicles to this day. An improved engine had a burn time of 12 minutes. These rocket engines can be seen at the National Museum of the United States Air Force at Wright-Patterson AFB, Dayton, Ohio.

The TEW16/43-13 was a rocket fighter from Arado, a low wing interceptor. It used the Walter HWK 509A engine, burning T-Stoff and C-Stoff. It carried automatic cannon in the nose. It was 9.7 meters long, with a wing span of 8.5 meters. Also from Arado was the E.381, "smallest fighter," not designed for pilot comfort. It was designed to be taken to altitude by the Ar234, an operational jet bomber. One of the bombers can be seen at the Smithsonian's Udvar-Hazy Center at Dulles Airport, Virginia. The smaller craft had a short powered flight and a long glide. The pilot flew prone to provides the plane with a low profile.

The Junkers EF-126 was supposed to be powered by a pulse jet engine, like the V-1 "cruise missile." It was converted to the rocket-powered EF-127. These were not produced by the end of World War-2.

The So344 was a strange and desperate design. The nose was a 400 kilogram explosive device, which was detachable, and had a proximity fuse. The plane was flown to altitude on another plane, and released to use its Walter HWK 109-509 rocket engine to enhance its speed. The pilot would release the payload into a formation of bombers, and then land. There was some question of pilot survive-ability. Only a scaled -down prototype was built.

Even Zeppelin wanted to be in the rocket plane business. They proposed the Fliegende Panzerfaust (Flying Armored Fist). It would be towed above the altitude of a bomber formation by a fighter plane, and released. It would then ignite its six solid fuel engines. After its mission, due to the rockets burning out, the center of gravity shifted to a bad location. The plane was to be designed to break in half, with each half parachuting to the ground. This one did not get off the ground.

The ME-262 was developed as an improvement to the -163, but only three prototypes were produced by the end of the war, and none flew under their own power. The JU-248 rocket plane was designed by Junkers. The Americans captured the plant in April of 1945, and secured the prototypes. One went to the U.S., and one to Russia. The Russians used it as a basis for their MiG-270 rocket power interceptor. This had a maximum speed of 581 mph, and could travel to more than 55,000 feet.

Another simpler rocket plane design was used in combat by the Germans. This was the BA-349 Natter. It used a liquid fueled Walter engine, and was launched vertically. After the mission, the rocket engine and the pilot parachuted to the ground, and the rest of the plane crashed. It was built to be expendable; only the engine and the pilot were important. The plane made its first manned flight in March, 1945, and killed the pilot.

The German V-1 was what we now call a cruise missile, and was operated unmanned. It used a pulsejet engine, which differs from both a jet engine and a rocket engine. A rocket engine carries its own supply of oxydizer. A jet engine uses atmospheric air, but must be moving forward. The pulse jet is a jet engine that develops thrust without moving forward. The V-1 accomplished this with a carefully selected diameter and length of the exhaust pipe, to achieve resonance. It was limited in performance, as there is almost no compression of the air. Also, pulsejets can't usually take off on their own, due to limited power production at low speeds. The V-1 used a launch catapult. A Russian inventor patented a pulsejet in 1864, but it doesn't seem to have been built. Another inventor produced a working engine in 1906. A pulsejet was built by Ramon Casanova in Spain in 1917. Robert Goddard did a pulsejet in 1931, and used it on a rocket powered bicycle.

Famed rocket scientist Werner von Braun contributed to rocket projects besides his V-1 and V-2. He supplied rocket motors to other experimenters, and he proposed a vertical take-off interceptor plane, launched vertically. Ir was initially planned to launch from a dedicated building holding multiple aircraft, later modified to being launched from a transport truck  It would have a pressurized cockpit, and was designed to glide back to a landing. It was to have a service ceiling of 8.000 meters, with a cruise speed of 690 km/hr. It was never built.

Japan

The Japanese received some rocket-plane help from Germany during World War-II. Mitsubishi was set to build ME-163's, but the Germans submarine carrying the airframe was lost to enemy action in the Atlantic. The submarine carrying the engines and detailed plans was sunk near the Philippines. A prototype was finally constructed in 1944, and flew successfully. Mitsubishi, Fuji, and Nissan collaborated on the airframe. The Japanese variant was lighter, although they had left out some of the armor. Sixty trainers and

7 operation craft were built. First powered flight was in July of 1945, but the plane crashed and killed the pilot. Full production was ramping up when the War ended. Four different rocket plane projects were ongoing, including a kamikaze version.

Soviet Union

Both the Soviets and the allies captured some of the German rocketplanes, and these units had an influence on post-war rocket planes. The Soviets found five Junkers 126 pulsejet prototypes in a factory in their zone of occupied Germany. Unfortunately, their pilot was killed during unpowered testing.

Soviet rocket plane development had kicked off in 1932, with Korolev's GIRD-6 Project. He developed the Rp-318-1 rocket plane that flew in 1940. Korolev went on to lead the Soviet rocket programs during the Cold War, and saw Cosmonauts flying to space on his rockets.

The Soviets developed their own rocket powered fighter planes, one being the Bereznyak-Isayev BI-1. This was a project of famed Soviet rocket designer Korolev. First flight was in 1942. It used an engine burning kerosine and liquid oxygen. The first two prototypes were built by furniture makers. It was flown 10 times under rocket power, with 2,430 lbf thrust. It could achieve 497 mph.

The earlier Cheranovsky RP-1 was a rocket powered glider, flown in 1933. It had some engine issues. The Chyeranovskii BICh-11 was designed around a twin engine cofiguration using the Tsander OR-2 engines. These used liquid oxygen and gasoline, and produced 1000 pounds-force of thrust. These were never certified for flight. The airframe fles as a glider, and later with a piston engine.

The Bereznyak-Isayev BI-1 was a war-time effort. There were problems with propellant and oxydizer feed, and compressed gas and turbo-pumps were tested. The plane was tested as a glider in 1941. Test flights were done in the relative safety of the Ural Mountains. First powered flight was in 1942. A series of test flights followed, culminating in a crash that killed the pilot. This was caused by lack of knowledge of aerodynamics at near trans-sonic speeds. Five more aircraft were built, and equipped with ramjet engines. Finally, rocket power was replaced by advancing turbojet technology, which was better understood.

The I-270n was a rocket powered interceptor aircraft. It was based on an earlier project, the RP-318, a 1936 design. Two prototypes of the I-270 were built and flown, and both were involved in crashes. The program was then discontinued. An earlier prototype during World War-II was nicknamed the "Devil's broomstick." The rocket plane work in the Soviet Union was based on the work of pioneers Tsiolkovsky and Korolev.

The MiG-105, *Lapot,* was a Russian crewed test vehicle to explore the domain of low speed landings. It was to be an orbital spaceplane. Work began in 1965, halted in

1969, but brought back again in 1974 to compete with the Space shuttle. It was discarded in favor of the Russian Buran.

Early U. S. and Allied effort

The first American crewed rocket flight had taken place in August of 1941, using a commercial airframe, the Ercoupe, and solid propellant rockets. The plane maintained its piston engine, but could take-off using either or both. This lead only to development of rocket-assisted take-off for heavy bombers. This was under the guidance of Dr. Theodore von Karman, The pilot was Army Air Force Captain Homer A. Boushey.

The next effort involved a futuristic looking flying wing design, the Northrop MX-324. This used a liquid fuel engine, but it was too underpowered to take off on its own. It was towed to altitude by a P-38 fighter. But, in 1944, the Army Air Force let a contract to Bell Aircraft for the first really successfully rocket powered aircraft built in the U.S. This was the X-1, which would go on to be the first to break the sound barrier. This would also kick off the military's X-plane Project, which covered all sorts of experimental aircraft.

The British put large solid fueled rockets on Hawker Hurricane piston aircraft, and these could take off from a small ramp installed on merchant ships. This provided protection against enemy submarines and aircraft. The plane could not be recovered unless they were close to land, and the pilot needed to parachute out.

The prototype of the Messerschmitt P.1101 was about 80% complete when the Americans captured it. It was taken back to the States and studied. It influenced the Bell X-5.

## Post war and Space Age

The development and maturation of the rocket airplane would lead to two milestones in the post-war era. The first was the breaking of the sound barrier by Chuck Yeager in 1947. Going trans-sonic removed a barrier to speed. The next limit to be broken was the ability to get to 100 km in altitude, beyond which was space. This was accomplished by the X-15 project.

In 1950, the Douglas D-558 Skyrocket was to be produced for the Navy as a "hybrid" with both jet and rocket motors. In this aircraft, Scott Crossfield flew at twice the speed of sound, shortly before the anniversary of the Wright Brother's first flight. By that time, the jet engine idea was abandoned, and a larger rocket engine was installed. There were three aircraft, and much data was acquired about stability in the transition to trans-sonic flight. This was useful for the development of the X-15, and the Space shuttle. The planes made over 300 flights, completing altitude and speed records. Unfortunately, these were not officially recognized, as the plane did not take off under its own power, but was carried to altitude by a B-29 mothership. These flight were conducted under NACA, the predecessor of NASA.

All three aircraft survived, and are on display at the

Planes of Fame Museum, in Chino, California; at the National Air and Space Museum in Washington D.C.; and on the grounds of Antelope Valley College, in Lancaster, California.

## U. S. Rocketplanes

This section discusses the United States' rocketplane efforts, Post-World War-II.

# X-1

The Bell X-1 was the first plane to exceed the speed of sound in controlled flight, controlled by Chuck Yeager. It was designed and built in 1945. The German war effort in 1944 brought rocket-powered fighter planes to production and deployment, but there was not much know about the transition region to faster-than-sound flight. England, starting in 1942, also began to develop a faster-than-sound aircraft.

The X-1 first flew in 1947. It was thirty feet long, with a wing area of 130 square feet. Empty weight was 7,000 pounds, 12,225 pounds loaded. Theoretical maximum speed was Mach 1.25. The engines could operate for 5 minutes on the supplied fuel, and then the X-1 became a very fast glider. It could reach 72,0000 feet, not quite space. It was carried to altitude by a B-29.

The X-1 used the XLR-11 engine from Reaction motors, producing 6,000 pounds-force from ethyl alcohol and liquid oxygen. It had four combustion chambers that could be operated independently, but were not capable of being throttled.

The X-2 Starbuster came out in 1955, a joint project by Bell, the U.S. Air Force, and NACA. It could go faster than the X-1 (Mach 3), and was used to investigate transsonic aerodynamic heating. It's engine, the XLR-25, was throttleable. It was dropped launched from a B-50 bomber. It's first, unpowered flight went well, but an in-flight explosion while still attached to the mothership killed the pilot and one crew member in the mothership. The first powered flight occurred in 1955. The first flight above Mach 3, a virtually unknown flight region, also killed the pilot. Exploration of this realm was postponed until the X-15 was ready for flight.

# X-24

The X-24A was a lifting body design. It was built by Martin Marietta, and taken to altitude by a B-52 mothership. It was used to prove concepts for unpowered landings later used by the Shuttle. First flight was in 1969, and continued through 1975. There was only one plane built, but it was extensively rebuilt by Martin, and renamed X-24B. This was a joint NASA-USAF Program.

The craft used the Reaction Motors XLR-11rs engine, capable of nearly 8,500 pounds of thrust. The A model was flown 28 times at speeds exceeding 1,000 mph, reaching 71,000 feet. There was a proposed X-24C which did not make it off the drawing board.

Similar aircraft from Northrop, the M2-F2, M2-F3, and HL-10 flew in 1966-67, with similar engines. These influenced the design of the Dreamchaser from Sierra Nevada. The HL-10 is on display at the entrance to the Armstrong Flight Research Center, at Edwards AFB, California.

# X-30

The X-30 was a Rockwell project for a technology demonstrator for the U.S. National Aero-Space Plane Project, a single stage to orbit design. The project was canceled in the 1990, without a prototype being produced. It did produce a lot of new knowledge in advanced materials. Some similarities with the British Skylon program are noted; the slushy hydrogen fuel was passed over the wings to cool them, taking that heat energy into the engines. The X-30 project, perhaps overly ambitious, was canceled due to budget cuts. There is a mockup at the U. S. Space Camp facility at the Marshall Space Flight Center in Huntsville, AL.

The plane was to be 160 feet long, with a wingspan of 74 feet. It would weigh over 300,000 pounds. It was a waverider design, using a scramjet engine. The scramjet (supersonic combusting ramjet), according to wikipedia, "is a variant of a ramjet airbreathing jet engine in which combustion takes place in supersonic airflow. As in ramjets, a scramjet relies on high vehicle speed to forcefully compress the incoming air before combustion (hence ramjet), but a ramjet decelerates the air to subsonic velocities before combustion, while airflow in a scramjet is supersonic throughout the entire engine. This allows the scramjet to operate efficiently at extremely high speeds." A waverider takes advantage of compression lift, produced by its own shockwave, at speeds above Mach 5. This enhances the craft's lift to drag ratio. The concept was introduced in 1951.

The NASA X-43 Hyper-X was an unmanned hypersonic aircraft. It is the fastest aircraft on record at Mach 9.6. The X-43 with a winged booster rocket was dropped from a B-52 aircraft. The booster rocket was a modified first stage of an Orbital Sciences Pegasus. The X-43 plane used a scramjet. Two flew successfully, and one was destroyed in flight. Engine burn time was 10 seconds. The project was conducted by NASA's Langley and Dryden Centers. Water had to be circulated around the engine cowling above Mach 3 so the aircraft would not melt. The fuel was hydrogen. The engine would only operate at speeds of Mach 4.5 or higher. The target speed for two of the units was Mach 7, where the other unit was targeted to Mach 9.8 or higher.

## Blackstar

The Blackstar program was a classified orbital spaceplane. It supposedly was operated by the U. S. National Reconnaissance Office, not the Air Force. It was a two stage to orbit vehicle, with a mothership called the SR-3, perhaps a modified North American B-70, and the rocket-powered payload, the XOV (experimental orbital vehicle). The XOV was similar to he X-20 Dynosoar, with supposedly an aerospike engine, which is a rocket without a nozzle. The XOV would land on a standard runway.

## HL-20

NASA's HL-20 "personal launch system" was studied at the Langley Research Center around 1990. It would use a standard rocket launcher, and be capable of landing on conventional runways. Langley preferred a lifting body approach, where the bottom of the vehicle was one big wing. The plane was towed flight d a C-47 in 1963. There were some 400 ground tows, and 77 aircraft tows. Besides the Shuttle, these planes influenced the Air Force's X-24 rocket plane. It was flown by a pilot to more than 74,000 feet, and achieved 1,165 mph. It can be seen at the National Museum of the Air Force at Wright-Patterson Air Force Base, Dayton, Ohio.

## X-15

The X-15 was a crewed hypersonic aircraft, designed to fly to the edge of space, defined as 100 km . A pilot who reached this altitude was officially an Astronaut. The X-15 was carried to altitude under the wing on a B-52 Mothership. After it was released, its rocket engine was ignited, pushing it up out of most of the atmosphere. There were no air breathing engines. It did a "dead-stick" landing. There were two flights to the 100 km altitude, both by Joseph A. Walker in 1963. The X-15 flights were

USAF, U.S. Navy, and NASA sponsored.

The X-15 holds the official world's record to the highest speed recorded by a manned, powered aircraf,t 4,250 mph (Mach 6.7) at 102,100 feet altitude. That was in 1967. It is officially the world's first space plane. It made further contributions to winged space flight, particularly the Shuttle Program, in the areas of weightlessness, and thermal protection for returning from extreme altitudes. The X-15 demonstrated a reaction control system that controlled the planes attitude in roll, pitch, and yaw when the conventional control services didn't have any air to push against. The X-15 pilots also tested new space suits.

The X-15 was developed from a concept by Dr. Walter Dornberger for NACA in 1954. He was one of the captured German scientists that formed the core of the rocket team. Dornberger served as the military officer in charge of the German rocket program, and came to the U.S. with von Braun.

The X-15 operated in two distinct domains, and had both aerodynamic control surfaces, and rocket thrusters. The plane included a pilot ejection seat, usable up to Mach 4, and 120,000 feet. The main engines were dual *Reaction Motors* XLR11 units, using alcohol and liquid oxygen to achieve a total of 16,000 pounds of thrust. Earlier, a single XLR11 pushed the Bell X-1 to be the first aircraft to exceed the speed of sound (Mach1). Later, the X-15 was fitted with the upgraded XLR99 for 57,000 pounds of thrust for 88 seconds. Essentially, the engines generated 1,000,000 horsepower. Over 175 flights were made in that configuration. Three X-15's made a total of

199 test flights, the last in 1968. Twelve pilots flew the planes, including future astronaut Neil Armstrong, who would go on to become the first man on the moon.

A later improvement had the Honeywell MH-96 Flight computer use the appropriate system, aerodynamic or reaction, with pilot input. This was an early fly-by-wire system, which is now common in almost all aircraft.

X-15's can be seen today at the National Air and Space Museum in Washington, D.C. and the National Museum of the United States Air Force, located at Wright-Patterson Air force Base in Dayton, Ohio.

The follow-on to the X-15 was to be the X-15B. Rather than riding a B-52 mothership, it would sit atop SM-64 a Navaho missile, and be launched vertically. The X-15B program was canceled when NACA was dissolved, and NASA took its place, right after the Soviet launch of the first satellite. NASA chose the capsule rather than the spaceplane format. The work on the X-15B was recast as the Dynasoar.

X-20 Dynasoar

Dynasoar was a USAF Project with a winged spacecraft. It was developed by Boeing as the X-20. The project started in 1957, and was canceled in 1963, just at the beginning of construction. The trend in spacecraft went to space capsules that had heat shields, and returned on a ballistic trajectory to a ground or water landing. Dynasoar was designed to reach Earth orbit with a single pilot, using a Titan launch vehicle. It had an equipment compartment behind the pilot, that could be used for payloads. A variant, the X-20X had a rear crew compartment that could hold 4. It was reusable, like the Shuttle would be. Both included a trans-stage at the rear for orbital maneuvering, that would be jettisoned before reentry. The plane could dip into the atmosphere and back to change its orbital inclination, without a large expenditure of fuel, which an orbiting spacecraft would need.

The project went back to Dr. Dornberger, from the German World War-II rocketry efforts. He had detailed knowledge of Eugene Sanger's Silbervogel Spaceplane project. The Dynasoar was modeled on that. Quite a few studies were done by the major U. S. Aerospace companies. This was to be a successor to the X-15 research vehicle. The contract to build the vehicle was awarded to Boeing. Later in the program, seven astronauts were chosen from NASA and the Air Force to fly the Dynasoar, including Neil Armstrong.

In retrospect, the program was canceled due to uncertainty over the booster to be used, and a lack of planning and clear goals. The program kept changing requirements, and thus no one agreed on what to build. The Dynasoar did influence the later Space Shuttle design and operations.

Dynasoar was to be some 35 feet long, with a 20 foot wingspan. Its empty weight was around 10,400 pounds. It was to be able to achieve 17,500 mph, and a 22,000 mile orbit.

Despite cancellation of the X-20, the affiliated research on spaceplanes influenced the much larger Space Shuttle. The final design also used delta wings for controlled landings. The Air Force continued research into lifting body rockets, in a program called Spacecraft Technology and Advanced Re-entry Tests (START). An SV-5 vehicle was flown from the Pacific Test Range (Vandenburg AFB) to a point some 4,400 miles downrange

## X-33 and X-34

The X-33 was Lockheed Martin's Unmanned demonstrator suborbital spaceplane for NASA. It was the technology demonstrator for the VentureStar orbital spaceplane. It was validating the concepts of a single stage to orbit, reusable launch vehicles. This project was canceled, and NASA chose to develop the Orion capsule instead of a new spaceplane.

Orbital Sciences developed the X-34, a pilot-less reusable launch vehicle with a spaceplane form factor. There were two demonstrators built, and submitted to towed and captive flight tests. The units are in storage at Edwards AFB.

## X-37

The USAF's X-37 Orbital Test Vehicle is launched on top of a booster, in a fairing, like a satellite payload to orbit. It is fairly small. Aerodynamic surfaces such as wings and a tail are not great for vertical flight. In 1999, this was a NASA Project, later transferred to the Air Force. There was a drop test of the vehicle at Edwards Air Force Base in 2006. It was launched to space in 2010. It spent 8 months in orbit, returning successfully, and validating the ceramic heat shield, and landing procedures. Since then, missions have flown from both the Kennedy Space Center, and Vandenburg Air Force Base, the latter used for polar orbits.

It flew to orbit I September, 2017, on a Space-X Falcon 9. The Falcon booster returned to the its landing site at Kennedy. It was in the path of Hurricane Irma, and was hastily towed to the processing facility.

It is fully autonomous in operation, not requiring a crew for orbital operations, or reentry and landing. It is some 29 feet long, with a wingspan of about 15 feet. The maximum take-off weight is 11,000 pounds. The payload bay is 7x4 feet. It is designed for up to 270 days in orbit.

The original intent was to carry the plane to orbit in the Shuttle's Cargo Bay. Since the Shuttle fleet was retired, it now rides a booster rocket to orbit, Delta IV or Atlas V. As a DoD Project, many of the aspects of the space plane are classified. It has been lifted by the Scaled Composites *White Knight* aircraft, for an atmospheric drop test. This was demonstrated after a data link failure. A specific Air Force variant was developed from the NASA X-37, called the X-37B. It is designed for an orbital stay time of up to 270 days. The plane can land on the Shuttle runway at Kennedy Space Center, and can also return to

Vandenburg. The plane achieves Mach 25 upon reentry.

Its Aerojet engine develops 6,500 pounds-force. There is now a "C" variation in planning with the capability of carrying up to six astronauts in a pressurized cargo compartment. There would be no onboard pilot. The Project was managed by the Marshall Space Flight Center in Huntsville, Ala.

# XP-79

The Northrop XP-79 was intended to be a rocket powered flying wing fighter aircraft. It was to use mono-ethyl aniline and red fuming nitric acid as fuel and oxidizer. Two prototypes were built. It was later re-engined with turbojets. A test pilot was killed, and the project was canceled.

# STS-Shuttle

Rocket planes did the early research for the Space Shuttle. A series of USAF and NASA efforts with the lifting body design influenced the Shuttle design and reentry profile. It was roughly the size and weight of a DC-9 jetliner.

The Space Transportation System (STS) was a crewed launch and recovery system for spacecraft, that used rocket propulsion to achieve orbit, and glided back to Earth to land on a run-way. A major advantage of the Shuttle system was, when it carried a spacecraft to orbit, it could check to see if it survived the harsh launch environment. If not, the Shuttle could bring it home. Perhaps its major achievement was to repair the Hubble Space Telescope in orbit over the course of several missions. The Shuttle was instrumental in assembling the International Space Station.

At launch, the STS consisted of the winged Shuttle vehicle, a large liquid fuel and oxidizer external tank, and two solid rocket boosters. The solid rocket casings were retrieved from the ocean, refurbished and reused. The external tanks were not recovered, and were targeted away from shipping lanes in the Pacific and Indian oceans.

There was a mock-up, a prototype, and five flight units. Two of the flight units were destroyed, one at launch, one at reentry, both with loss of crew.

The Shuttle Orbiter rode the side of a large fuel and oxidizer tank to orbit, assisted by fall-away solid boosters to get everything going. The Shuttle Orbiter had three engines, fed from the large external tank. When the engines had burned sufficiently to achieve orbit, the Orbiter separated from the tank. The Orbiter continued to its destination altitude. The engines went with it, but no longer had a source of fuel or oxidizer. The Orbiter could adjust its orbit somewhat with its OMS (orbital maneuvering system) engines, using fuel onboard. There were also (reaction control system) RCS engines to adjust attitude. Upon reentry, the Shuttle flew in a nose-up attitude, as the bottom of the craft and wings were covered in heat-resistant tiles. After sufficient atmosphere was reached, the aircraft control surfaces could be used, and the Orbiter was flown like a plane to a runway landing. Well, like a 165,000 lb glider. No air-breathing engines were included. These were considered and rejected in the early design phase. They would have allowed the Shuttle to maneuver in the atmosphere and extend its range, at the cost of complexity and weight.

Ideally, the Orbiter landed at the runway back at the Launch site, and could be easily towed to the maintenance facility. Another option was to use the Dryden flight facilities vast expanses of hard desert. In that case, the Shuttle was brought back to the Kennedy Space Center on the back of a specially modified 747 carrier aircraft.

There was a plan to launch Shuttles from Vandenberg Air Force Base in California, which would allow them to go to polar orbit. This was not implemented. There was also a two stage rocket called the Interim Upper Stage or Inertial Upper Stage (IUS) which would deploy from the Shuttle bay with a payload going to a higher orbit. One mission carried the Ulysses spacecraft to study the polar regions of the Sun. Quite a few were used to put the Tracking and Data Relay Satellites (TDRS) into orbit.

A versatile craft, the Shuttle could take satellites to orbit, check them out, release them if they worked fine or bring them back, if they didn't. The Shuttles also made several repair trips to the Hubble Space Telescope, to work around its optical problem, and change out a failed computer. The shuttle carried up to 8 crew, some upstairs, some on the lower deck (cheap seats, no view), and could accommodate 11 in an emergency.

NASA re-purposed the Apollo Vertical Assembly Building (VAB) at the Cape to assemble the Shuttle stack. The solid boosters were bolted down on the crawler/transporter base, and the large external tank and Shuttle Orbiter were hoisted up and attached. This facility has such a massive volume inside, it has its own weather, There are two launch pads, 39-A and 39-B, essentially identical.

The launch sequence proceeded in a well defined procedure. The three liquid engines were ignited one at a time in sequence, to check that they were all working properly. This pushed the orbiter's nose forward about a meter. When the liquid engine performance was verified, the explosive bolts holding the solid boosters to the pad were blown, and the solid boosters were ignited. Then, you were on your way.

In some video of a Shuttle launch, you will see a series of sparks below the engines. That was to ignite any residual or leaked hydrogen from the external tank. A huge water spray was started before engine ignition. This was to partially protect the pad, but also to damp the acoustic energy from the engines. Otherwise it reflected up onto the vehicle, and could do damage.

The Space Shuttles carried five identical computers, the circa-1972 AP-101's, derived from the IBM System/360 System/4 Pi mainframe architecture. It was a 32-bit machine with 16 registers, and was microprogrammed. It had an instruction set of 154 opcodes. One of the five AP-101's on the Shuttle contained software derived independently from the software loaded on the other four. Each unit had a CPU and an IOP - Input/Output Processor. Each IOP had 24 channels, each with its own bus and processor. Triple redundant power supplies, fed by separate essential electrical buses were used. The computers were located in three separate locations in the Shuttle Orbiter. Redundancy is everything.

Want to see Shuttle flight hardware? *Pathfinder*, a full-size mockup, is at the Alabama Space and Rocket Center, Huntsville, AL. OV-101 *Enterprise*, a prototype used for flight tests in the atmosphere, is at the Intrepid Sea, Air & Space Museum in New York City. OV-102 *Columbia* was destroyed (with loss of crew) in a re-entry accident on February 1, 2003. OV-099 *Challenger* was *d*estroyed (with loss of crew) in a launch accident, January 28, 1986. Debris was recovered and is stored, sealed in an old missile silo, at Cape Canaveral Air Station, FL. OV-103 *Discovery* rests in the National Air and Space Museum, Steven F. Udvar-Hazy Center, Chantilly, VA. (near Dulles Airport). OV-104 *Atlantis* may be see at the Kennedy Space Center, Cape Canaveral, FL. OV-105 Endeavor is at the Samuel Oschin Pavilion of the California Science Center in Los Angeles, CA.

No flown external tanks have survived, but unused ET-94 is in Los Angeles and will be on display with Space Shuttle Endeavor at the California Science Center, when the Samuel Oschin Air and Space Center opens in 2018. Three external tanks were in manufacturing when the Shuttle Program ended, numbers 139-141.

One of the Shuttle Carrier aircraft, a specially adapted Boeing 747, can be seen at Palmdale (CA)'s Joe Davis Heritage Airpark. NASA retains ownership of the aircraft. The other carrier aircraft was placed at Space Center Houston, with a Shuttle mockup on top.

## X-38 Crew Return Vehicle

The X-38 was conceived as an emergency crew return vehicle for the ISS. The population of the ISS depends on how many human-rated craft are attached. This is currently 6, since there are two Soyuz capsules docked at any given time. When a new crew goes up, they return on the capsule that has been docked to the station the longest.

The X-38 had many innovative features. It could be flown to a safe landing by controllers on the ground. A para-sail was considered for landing. It was a lifting-body architecture

A lot of the vehicle was designed with COTS components. The thermal tiles were similar to those used on the Shuttle. The flight computers came from commercial aircraft. The Navigation system was in use on Navy fighters. It could also use GPS navigation.

Scaled Composites got a contract in 1996 to build three full-scale airframes for testing. The first was delivered that year. Drop tests were conducted from a B-52 aircraft at 45,000 feet altitude. Near-trans-sonic speeds were achieved. A drogue parachute slowed this to 60 miles per hour. Flight control was autonomous, with a stand-by pilot/controller on the ground.

NASA also partnered with the European Space Agency and the German Space Agency on this project. In usage, the X-38 would be carried to the ISS onboard the Shuttle. One rocket plane as cargo for the other.

As well as things were going, cost-overruns caused the cancellation of the program in 2002. The three flight test models can be seen at the Strategic Air and Space Museum, Ashland Nebraska; outside of Building 49 at Johnson Space Center; and at Evergreen Aviation Museum in McMinnville, Oregon.

## Other Nation's Spaceplane Projects

This section discusses the Spaceplane projects of other Nations, and is followed by a section of commercial ventures.

# Buran – the Russian Shuttle

Buran ("snowstorm" in Russian) was a Russian spaceplane, and the name of the entire project. It completed one 3 hour+ non-crewed flight (2 orbits) in 1988. It was commanded by ground control to de-orbit and landed at the runway at the Baikonur Cosmodrone in a strong cross-wind. It was tragically destroyed in its hanger, when the roof collapsed due to snow load in 2002. The collapse also killed eight workers. It resembled the U. S. Space shuttle, but included the auto-land mode. Buran had a sophisticated system of dual 4-computer units, with voting logic.

Like the Shuttle, the Buran was transported to the launch site on the back of a large aircraft, the Antonov AN-225 *Mriya (*Russian: "Inspiration*")*. That plane is the longest and heaviest ever built, and uses six turbofan engines. The maximum take-off weight is 640 metric tons. A second aircraft was partially constructed, and later finished for delivery to China. The aircraft has 32 wheels on its landing gear, and a crew of six.

Ref: http://www.antonov.com/aircraft/transport-aircraft/an-225-mriya

# Russian BOR-4

This vertical take-off, horizontal landing un-crewed plane was used to test the tiles for the Buran Shuttle. In the period of 1982-84, four test flights were made. Seven craft were built.

The Chinese had a spaceplane called the CSSHQ. It was firast launched in 2020. It is supposed to be similar to the Boeing X-37B.

# British Hotol

The Horizontal Take Off and Landing Spaceplane was a 1980's single-stage-to-orbit reusable craft with an airbreathing jet engine. It used ambient air as it climbed up from its runway, switching to an onboard supply of liquid oxygen at an altitude of 23-30 km. The craft was unmanned. It was supposed to be capable of putting 7-8 (metric) tons to 300 km. It was 62 meters in length, with a wingspan of 19.7 meters. It's wing design was based on the Concorde's. Reentry heating on the wing was estimated to be 1,400 degrees C, which was manageable. The Hotol project ended in 1989. The main technical leaders formed a new company, Reaction Engines, Ltd, and continued development of the key element, the "hybrid" engine.

# Skylon

Skylon is a single stage to orbit spaceplane using a hybrid air-breathing rocket engine, with hydrogen for fuel. It requires a special runway. It is sized for 37,000 pounds of cargo to low Earth orbit. Design goals call for 2-day turn-around, and a 200 flight life. The plane may be ready for operation in 2025. It is based on the previous HOTOL Concept from the 1980's.

The hybrid engine would burn hydrogen, using atmospheric oxygen until reaching around 26 kilometers, where it would switch to the internal liquid oxygen. The engine is referred to as *Sabre*, Synergetic Air-Breathing

Rocket Engine. It has been in development for over two decades.

It was designed to carry 37,000 pounds to LEO, or 24,000 pounds to the ISS. It will have retractable undercarriage, like the Shuttle. The plane is to be about 83 meters long with a wing span of 26.8 meters. It can be configured to hold 24 passengers in the Skylon Personnel Logistics Module. The plane would normally not have a Captain or Pilot, but would be flown from the ground.

The project is being conducted by the U.K. Company REL (Reaction Engines, Ltd.). The intent is to operate for-profit, since he Skylon has the potential to lower launch costs. However, they estimate that $12 Billion would be needed to develop an operational vehicle. The European Space Agency and the British Government are contributing funds. Commercial entity BAE Systems acquired a 20% share of the company.

Besides the orbital space plane, the same approach could be used to implement Sanger's antipodal airliner concept. This could be used in a peaceful fashion as a very fast passenger and freight delivery aircraft.

# German Sanger II

This was a proposed two stage to orbit winged spaceplane project in the 1980's, by the German Hypersonics Programme. It was to mass 366,000 kg, and have a thrust of 4,500 kn. Studies were done at

Messerschmidt-Bolekow-Bloehm. The first stage used a horizontal take-off, using a turbo-ramjet. The engine was built, and ground tested. The conclusion was that it would be very costly to develop, and showed minor cost advantages over the Ariane launch vehicle operationally. The project was canceled because of this.

The Horus winged winged second stage would have used a liquid oxygen/liquid hydrogen engine, with a thrust of 1,280 kilo-newtons. It was to carry 3,000 kg and two crew, or no crew and 15,000 kg.

## Indian Avatar

The Aerobic Vehicle for Trans-atmospheric Hypersonic Aerospace Transportation, *AVATAR*, is an Indian concept study for an unmanned reusable spaceplane. It would utilize horizontal take-off and landing from conventional airfields. It would use liquid hydrogen-liquid oxygen, some of the oxygen being collected on the upbound trip. This, it is partially air-breathing. This project is under the Indian Defense Research organization.

In Indian mythology, an Avatar is a descent of a deity to Earth. The Avatar Project involves a spaceplane that can take off from regular airfields It would operate in a similar fashion to the Skylon Project, who engines can be operated as jets, and rockets. The projected lift-off weight is 25 tons, 60% of that cyrogenic liquid hydrogen fuel. Oxygen would be collected and liquidized on the ascent. The plane is to achieve 1000 kg to low Earth orbit, and be reusable for 100 flights.

## French Hermes

The Hermes spaceplane was a proposed French project circa 1975. It was later taken over by the European Space Agency. It was to have a crew of six. The Shuttle Challenger accident drove a requirement for crew ejection seats, which limited the crew to three. Cargo capacity to orbit would be around 3,000 kg. The launch vehicle was to be the Ariane-V.

The project was discontinued due to budget reasons, and a perceived view that both the U.S. and Russia could provide launch services that the spaceplane was being designed for.

## ESA SOAR

The SOAR spaceplane was derived from the Hermes design, and was to be carried to altitude on an Airbus A300 Mothership. It was a project of Swiss Space Systems. It used a Russian engine, the NK-39, burning RP-1 and LOX. The company ran into financial difficulties, and went bankrupt.

## Airbus Defense and Space Spaceplane

This project involves a suborbital spaceplane for space tourists. It will have turbofan jet engines, and a methane oxygen rocket engines. It is sized for a pilot and four passengers. Airbus also patented a plane with turbojetx, ramjets, and a rocket engine.

## Commercial Ventures

This section discusses commercial rocketplane ventures. All of these (in the United States) require approval from the FAA's Office of Commercial Space Transportation. That office also licenses space ports, such as the Mojave Air & Space Port. This is in accordance with international law, governing rocket launches.

The EZ-Rocket research and test airplane was first flown in 2001. EZ-Rocket was the first privately built and flown rocket-powered airplane.

# Virgin Galactic

Virgin Galactic is providing suborbital flights, and plans to address space tourism as well as science missions. The company is owned by Sir Richard Branson.

Scaled Composites, owned by Virgin Galactic, developed a air-launched vehicle, for access to space for tourists. The carrier vehicle, White Knight Two, carries the smaller SpaceShipTwo to a high altitude,. At this point the two vehicles separate, and the White Knight flys back to a runway landing as the smaller craft continues to orbit with its rocket engine. The two vehicles were designed and built by aviation and space pioneer Burt Rutan. The plan is to have five SpaceShip-2's, the orbital vehicle, and two WhiteKnight-2's, the crewed carrier vehicle.

Two SpaceShip-2's are under construction, and there is one White Knight-2, the carrier ship.

The White Knight, sometimes called a Flying Space Aircraft Carrier, carry's a flight crew of two. It's cargo capacity is 37,000 lbs to 50,000 feet. If it carries an "upper stage", the Launcher One, it can put 200 kilograms to low Earth Orbit. White Knight has four Pratt & Whitney turbofan engines. The wingspan is 41 feet, and there are dual fuselages. The second fuselage can hold additional crew members, or tourists.

Unfortunately, the first SpaceShipTwo broke up during a test flight, killing one pilot, and injuring the other, although he parachuted to the ground. Before the accident, the vehicle did indeed reach "space." The second unit is undergoing flight testing at this writing. The first model was named *VSS Enterprise*, with a tail number of N339SS. The second is Unity, with a tail number of N202VG. There have been a total of 54 test flights at the writing.

It is designed to carry six passengers and two pilots. It can officially reach "space," defined as 100km. It glides back to a landing at Spaceport America in the Mojave Desert.

Virgin Galactic said if the Sub-orbital SpaceShip Three is successful, it will make the follow-on SpaceShipThree an orbital craft, and also investigate a point-point suborbital passenger and cargo delivery service.

# Sierra Nevada's Dreamchaser

This craft is a 7-crew Earth orbital space plane, developed with $20 million of NASA seed money. The Dream Chaser was based on the circa-1990 NASA HL-20 Personal Launch System, a mini-Space Shuttle designed as a lifting body. NASA-Langley, the lead center for the project, had not gotten to the implementation stage. The unit would have limited cargo capacity, but could get crews to and from orbit, particularly the ISS. The Company took part in NASA's Commercial Crew Development Program.

Sierra Nevada did not make the final cut, losing to SpaceX and Boeing, supposedly due to lack of technology maturity. Sierra Nevada decided to pursue the Dream Chaser project independently. It was not selected for the crewed vehicle contract.

Sierra Nevada did win a Commercial Resupply Services contract from NASA, for a minimum of six launches . These missions used the Atlas V launch vehicle.

Sierra Nevada also supplied the rocket motors for Virgin Galactic's Spaceship Two, and for Scaled Composites's Spaceship 1 and 2.

## Scaled Composites

This company was founded by famed aircraft designer Burt Rutan, and currently owned by Northrop Grumman. It is located at the Mojave Spaceport in California. The company is known for its unconventional designs, and use of advanced composite materials. It was granted the world's first license for a sub-orbital manned rocket flight.

SpaceShipOne was designed to win the Ansari X-Prize, for the first private manned spaceship. In 2004, SS1 made 3 crewed suborbital flights into space (above Kaman line), being the first privately built and operated craft to do so. The craft went on to win the X-Prize in 2004.

Spaceship one (Scaled Composites model 316) is an air-launched rocket powered aircraft. It made its maiden flight on the 100$^{th}$ anniversary of the Wright Brother's flight. It carries a pilot and two crew or passengers. It has a reaction control systems (jets) for attitude control outside the atmosphere, as well as standard aerodynamic control surfaces. The wings can be tilted forward to form a shuttlecock configuration, useful in the early stages of reentry. The craft cannot take off on its own, and is carried to altitude by the White Knight mothership.

The craft uses a hybrid rocket engine from SpaceDev, with a solid, rubber-like propellant, and nitrous oxide. The engine can be shut down after it is started, but it is not throttleable. It has a total burn time of around 80 seconds. The craft is 28 feet long, with a wing span of 16 feet, 5 inches. It's empty weight is 2400 pounds. SpaceShipOne has flown multiple times.

SpaceShip Two (Model 339) is a larger model, holding 8 crew and passengers in total. It uses a larger rocket model, and the same feathered or shuttlecock stabilization for atmospheric reentry. It is roughly twice the size of its predecessor. In a tragic accident, the first model broke up in flight, and crashed into the desert, killing the pilot. A second unit has been constructed.

The White Knight One (Model 318) is the carrier aircraft for SpaceShip-1. It takes off and lands from a normal runway, and carries the SpaceShip to an altitude of 15 kilometers. It uses twin turbojet engines. The combination of planes can land safely together.

White Knight Two (Model 348) is the follow-on mothership, being three times larger than the original. It is used with the SpaceshipTwo. It has four jet engines. Virgin Galactic has ordered two. They want to order 3 additional, plus 5 of the SpaceShipTwo models, giving them an orbital fleet. It is, in effect, a jet powered first stage.

Scaled Composites manufacturers the Model 351 to use as a mothership. It happens to have the longest wingspan of any aircraft to date, at 385 feet.

Michael Melvill, on June 21, 2004, went to a height of 100 km (62 mi), making him the first commercial astronaut. The prize-winning flight was flown by Brian Binnie, which reached a height of 112.0 km ( breaking the X-15 record.)

A joint venture between Scaled Composites and Virgin Galactic is named, "the Spaceship Company." As of 2012, Virgin owns the company in total.

# Rocketplane Kistler

Rocketplane Kistler was formed in 2006 by the merger of

Rocketplane, Ltd and Kistler Aerospace in Oklahoma. They had a NASA contract, but failed to meet financial milestones, and filed for bankruptcy in 2010. It's planned K-1 rocket was a reusable launch system. The assets of the company were purchased by Rocketplane, Limited, of Wisconsin. That company was in and out of bankruptcy. The concept is a that takes off with jet engines and climbs to altitude, where it loads fuel and oxydizer from a tanker aircraft. It then goes up to 100km, where a second stage is released to orbit. The project name was Rocketplane XP

New companies emerged from the debris, including Space Assets, Ltd and Kistler Space Systems. Rocketplane Global is a Wisconsin-based company, formed from Rocketplane, Ltd after bankruptcy.

# XCOR

XCOR is based at the Mojave Air & Space Port in California, and the Midland International Air and Spaceport in Texas.

The XCOR EZ-Rocket is a test platform, built from a Rutan Long-EX aircraft, with the propeller and engine removed, and a pair of liquid fueled rocket engines added. It was first flown by Rutan in 2001. The engines are rated at 400 pounds-force each. It can climb at 10,000 feet per minute. The engines use alcohol and liquid oxygen.

The Lynx spaceplane is a sub-orbital winged craft that will reach 100 kilometers altitude. This will officially get you astronaut status. The company booked 175 flights by 2012, at a cost of $95,000 each. A similar design was done for the Rocket Racing League. This was the Mark-I racer.

The Rocket Racing League, sometimes called NASCAR with Rockets, was founded in 2005. Three prototype craft were produced, based on the XCOR-modified Velocity SE, and the Armadillo Aerospace Velocity XL. By 2008, there were 6 teams registered in the league. The program never got off the ground, and the league was defunct by 2014.

The Lynx spaceplane project has been placed on hold, while XCOR concentrates on development of a new rocket engine. They are also working on a liquid hydrogen-liquid oxygen engine with United Launch Alliance.

## Orbital Sciences Prometheus

The Prometheus space plane would use a vertical take-off, horizontal landing scenario. It was targeted to NASA's Commercial Crew Development Program. It was based on the earlier NASA HL-20 Orbital Space Plane Program. It was to carry four astronauts to the ISS. NASA did not select the Prometheus project for funding, and Orbital declined to pursue the effort as a commercial venture. The requirement was to carry 4 crew. It had to provice crew return in case of an emergency, and serve as a 24 hour safe haven.

The first use of the program was to develop vehicles to take and remove crew.Space-X and Boeing took the lead. Boeing's CST-100 was also in the game. Others that participated include Paragon, Sierra Nevada, United Launch Alliance, Paragon Space Development, and I/Space.

## Bristol Spaceplanes

Bristol Spaceplanes is a U. K company designing planes with sub-orbital and orbital capability. They want to address the cost of access to space by providing re-usable launch systems, The company was founded in 1991. Their initial design studies were for the Ascender and Spacecab. The Ascender is a two-seat suborbital craft that does horizontal take-off and landing, It is both jet and rocket powered, and designed to get to 100 km, which gets you the official title of Astronaut. It would achieve 2 minutes of weightlessness at the apex of the flight.

In a typical profile, the plane would take off from a conventional runway using turbofan engines. At cruising altitude, it would ignite its hydrogen peroxide rocket engine, and climb to 80 kilometers or so, then return to a runway landing. The jet engines can be re-ignited, and used for the landing.

Spacecab is a project built around two aircraft, a large

carrier, and a smaller orbiter. The large craft would, like Ascender, have both jet and rocket engines. The orbiter would have cyrogenic rocket engines to deliver 750 kg to orbit.

Spacebus is a scaled up Spacecab. The mothership would have both ramjet and rocket engines, and would achieve Mach 6. The smaller vehicle would use rocket engines to achieve orbit with a 5,000 kg payload, or 50 passengers.

## Wrap-up

Since the early days, aircraft pioneers have sought to go higher and faster. At the same time, development in rocket engines showed these to be superior to propeller craft, and even air-breathing jet engines. This all came together in Germany in world War-II, in response to the massive Allied bombing raids. After the War, the Allies used the prototypes to develop their own rocketplanes. One of the first things to explore was the strange transsonic region, where controls could reverse, and airflow separated from winds and control surfaces. Without a group of dedicated test pilot-engineers, who put their life on the line with every flight, this would not be know.

As we ventured into space, the main recovery method was the capsule, using heat shields and parachutes. Based on the rocket plane research, the Space Shuttle was developed, which took off like a rocket, and landed like an airplane.

Now, several commercial entries use the method of a jet-power motherplane to take a smaller rocket plane to altitude, then release it to continue to orbit. The smaller plane can re-enter and land on a runway. The re-usablility of this system leads to economy of scale.

A re-examination of Sanger's rocket Plane concepts for speeding up long air journeys for passengers and cargo is being conducted.

Next, we need to know what aerodynamic vehicles we can use to explore Venus, Mars, the Gas Giants, and their many moons.

# Bibliography

Anderson, John *X-15: The World's Fastest Rocket Plane and the Pilots Who Ushered in the Space Age*, Zenith Press, 2014, ISBN-0760344450.

Belfiore, Michael *Rocketeers, How a Visionary Band of Business Leaders, Engineers and Pilots is boldly Privatizing Space,* 2008, Collins, ISBN 978-0-06-114902-3.

Bentley, Matthew A. *Spaceplanes: From Airport to Spaceport,* 2009, Springer, ASIN-B008BB7HQA.

Bredt, Irene, Sanger, Eugen *A Rocket Drive for Long Range Bombers*, 1944, reprinted 1952 by U. S. Government, Translation CGD-32, ASIN-B073RVFHXS.

Brodsky, Robert *Catch a Rocket Plane,* 2012, Foxbro Press, ASIN-B009Z51NN4.

Buss, Jared S. *Willy Ley: Prophet of the Space Age*, 2017, University Press of Florida, ISBN-0813054435.

Caiden, Martin *X-15: Man's first flight into space,* 1959, Scholastic Book Services, ASIN-B0007FSAW8.

Caiden, Martin *Wings into Space: The History and Future of Winged Space Flight,* 1964, Holt, Rinehart and Winston, ASIN-B000QLYCTI.

Christopher, John T*he Race for Hitler's X-Planes*. The Mill, Gloucestershire: History Press, 2013. ISBN 978-0-7524-6457-2.

Crossfield, A. Scott, and Blair, Clay *Always Another Dawn: The Story of a Rocket Test Pilot,* 1960, World Publishing Co. ASIN-B0010OYBXM.

DARPA/NASA "Report on the Horizontal Launch Study," June 2011, Interim Report, avail:
 https://ntrs.nasa.gov/archive/nasa/casi.ntrs.nasa.gov/20110015353.pdf.

Deidrich, Hans-Peter *German Rocket Fighters of World War II (Schiffer Military History)*, 2005, Schiffer Publishing, ISBN-0764322206.

DoD *High Frontier: The U. S. Air Force and the Military Space Program - MOL, Dyna-Soar,* 2017, ISBN-1520775709.

Dornberger, Walter R. "The Rocket-Propelled Commercial Airliner" Dyna-Soar: Hypersonic Strategic Weapons System, Research Report No 135. University of Minnesota, Institute of Technology, 1956.

Evans, Michelle *The X-15 Rocket Plane: Flying the First Wings into Space*, 2013, U. Nebraska Press, ASIN-B00CGVLH0S.

Forsyth, Robert (Author), Laurier, Jim (Illustrator),

*Messerschmitt Me 264 Amerika Bomber* (X-Planes), 2008, ASIN-B01N6CD4WQ.

Franks, Richard A. *The Messerschmitt Me 163: A Detailed Guide to the Luftwaffe's Rocket-Powered Interceptor,* 2016, Valiant Wings Publishing, ISBN-0993534538.

Goodwin, Robert *Dyna-Soar: Hypersonic Strategic Weapons System,* Apogee Books Space Series 35, ISBN-1896522955.

Gorden, Zefim, *Soviet Rocket Powered Fighters* (Red Star), 2006.

Green, William, *Rocket Fighter (German Komet fighter),* 1971, 1$^{st}$ ed, Ballantine Books, ISBN-0345258932.

Griehl, Manfred *German Rocket Planes,* 2000, Greenhill Books, ISBN-1853674044.

Hallion, Richard P.; Gorn, Michael H. *On the Frontier: Experimental Flight at NASA Dryden*, 2003, Smithsonian, ISBN-1588341348.

Hendricks, Bart *Energiya-Buran: The Soviet Space Shuttle,* Springer Praxis Books, 2007, ISBN-0387698485.

Hendrickson, Walter B. *Winging Into Space*, 1965, Bobbs Merrill, ASIN-B001N5YOL2.

Jenkins, Dennis R., Lands, Tony R. *Hypersonic: The Story of the North American X-15*, 2003, Specialty Press, ISBN-1580071317.

Kubal, Quinton *Spaceplanes: The Future of Space Travel*, 2015, CreateSpace Independent Publishing Platform, ISBN-10-1519183577.

Lardas, Mark Palmer, Ian *Space Shuttle Launch System 1972–2004* (New Vanguard), 2012, Osprey Publishing, ASIN-B01DPPWI74.

Launius, Roger D.; Jenkins, Dennis R. *Coming Home: Reentry and Recovery From Space,* 2012, ISBN-0160910641.

Ley, Willy, Rockets, *The Future of Travel Beyond the Stratosphere*, 1945, Viking Press, ASIN- 0007E7IC2.

Ley, Willy *Rockets, Missiles & Space Travel*, 1958, Viking Press, reprinted Signet,1969, ASIN-B001Q9KTGE.

Ley, Willy; Polgreen, John (Illustrator) *Space Pilots* (Adventure in Space), 1957, Guild Press; 1st, ISBN-1122700733.

Libis, Scott *Douglas D-558-2 Skyrocket* (Naval Fighters N0. 57), 2002 Ginter Books, ISBN 0-942612-57-4.

Miller, Jay and Yeager, Chuck, *The X Planes: X-1 to X-*

*31,* Aerofax Inc.; Revised edition, 1988, ISBN-0517567490.

Mullane, Mike *Riding Rockets: The Outrageous Tales of a Space Shuttle Astronaut,* 2006, ISBN-0743276825.

Myhra PhD, David *Heinkel He 176-Redeaux,* 2013, ASIN-B004H4XNJK.

NASA, "Proceedings of the X-15 First Flight 30th Anniversary Celebration - NASA Conference Publication 3105 - Historic Hypersonic Rocket-powered Spaceplane and Aircraft Tales, X-30 and Space Shuttle" 2015, ASIN-B00YSJOOGU.

Oborny, Lester F. et al, *Introduction to the Problem of Rocket-Powered Aircraft Performance,* 1947, NACA-Langley, ASIN-B00CJIXD3C.

Periscope Film.com *X-15 Rocket Plane Pilot's Flight Operating Manual,* 2006, ISBN-10-141169824X.

Presidential Commission (Rogers Commission), *Report on the Space Shuttle Challenger Accident, 1986,* US Government, 5 volumes, Vol. 1 – main report, ISBN-999739769X.

Ransom, Stephen *Jagdgeschwader 400: Germany's Elite Rocket Fighters,* Osprey, 2010, ISBN-1846039754.

Ransom, Stephen *Me 163: Rocket Interceptor,* 2022, ISBN-978-1906537579.

Redding, Jr. Fred W. *Spaceplane Technology and Research (STAR)*, 1984, ASIN-B00CC26S0K.

Sanger, Eugen *Space Flight, Countdown for the Future*, McGraw Hill; First U.S. Edition (1965), ASIN-B003VZRI6A.

Sanger, Eugen, "Techniques of Rocket Flight," 1933, privately published.

Sänger, Eugen; Sänger-Bredt, Irene "A Rocket Drive For Long Range Bombers", 1944, available:http://www.astronautix.com/data/saenger.pdf

Sänger, Eugen (Author), Fruch, Karl (Translator) *Space Flight: Countdown for the Future,* 1965, McGraw Hill 1965 ASIN-B0018SOIK6.

Sanger, Eugen *Rocket Flight Engineering* (NASA technical translation), NASA TT F-223, 1965, ASIN-B0007EYP2I.

Seedhouse, Erik *XCOR, Developing the Next Generation Spaceplane* (Springer Praxis Books), 2016, ISBN-331926110X.

Seedhouse, Erik *Virgin Galactic: The First Ten Years* (Springer Praxis Books), 2015, ASIN-B00TY7N7F0.

Seedhouse, Erik *SpaceX's Dragon: America's Next Generation Spacecraft,* 2016, Springer, ISBN-

3319215140.

Seedhouse, Erik *Bigelow Aerospace: Colonizing Space One Module at a Time*, 2015, Springer, ISBN-3319051962.

Sivolella, Davide *The Space Shuttle Program: Technologies and Accomplishments* (Springer Praxis Books), 2017, ISBN-13319549448.

Sparks, Major James C. *Winged Rocketry,* 1968, Dodd Mead, ISBN-0396057799.

Stack, John; von Dornhoff, Albert E. *Tests of 16 Related Airfoils at High Speeds*, NACA, circa 1937, Report 492.

Stone, Robert *How to Become a Virgin Galactic Space Pilot*, 2012, ISBN-1469948745.

Thompson, Milton O.; Armstrong, Neil A. *At the Edge of Space: The X-15 Flight Program,* 2013, Smithsonian Books, ASIN-B00DFIDW9K.

Tobin David M. *Man's Place in Spaceplane Flight Operations: Cockpit, Cargo Bay, or Control Room?* 2012, ASIN-B012UMMO0Y.

Gordon, Yefim, 2006. *Soviet Rocket Fighters*. Hinkley: Midland. pp. 69–84. ISBN 978-1-85780-245-0.

U.S. Air Force, *Bell X-1A Rocket Plane Pilot's Flight*

*Operating Instructions*, ISBN-1430308079.

U.S. GAO Space Shuttle Accident, NASA's Actions to Address the Presidential Commission Report, October, 1987, GAO/NSIAD-88-30BR.

U. S. Government *The Rise and Fall of Dyna-Soar: A History of Air Force Hypersonic R&D, 1944-1963 - Pathfinding Effort to Develop a Transatmospheric Boost Glider and Spaceplane, Manned Military Space Program,* Progressive Management, 2015, ASIN-B00VJK0NNY.

U. S. Government, *X-15: Extending the Frontiers of Flight - Encyclopedic History of America's First Hypersonic Rocket-powered Aircraft and Space Plane,* 2015, ASIN-B00YSITUEM.

van den Abeelen, Luc *Spaceplane HERMES: Europe's Dream of Independent Manned Spaceflight*, 2016, Springer 1st ed, ISBN-3319444700.

Vance, Ashlee *Elon Musk: Tesla, SpaceX, and the Quest for a Fantastic Future*, 2015, Ecco, ISBN-0062301233.

Van Pelt, Michel *Rocketing Into the Future: The History and Technology of Rocket Planes,* 2012, Springer Praxis Books, ISBN-1461431999.

Warstiz, Lutz *The First Jet Pilot: The Story of German Test Pilot Erich Warsitz,* 2009, Pen & Sword, ASIN-

B00AE7DHFY.

Williams, Walter C. *Flight research at high altitudes and high speeds with rocket-propelled research airplanes,* 1955, SAE, ASIN-B0007EJN1Q.

Ziegler, Mano *Rocket Fighter: The Story of the Messerschmitt Me 163,* Doubleday, 1961, ASIN-B000H6D3Q0.

Ziegler, Mano *Messerschmitt Me 163 Komet,* Schiffer Publishing Ltd, 1990, ISBN-0887402321.

## Resources

- https://www.grc.nasa.gov/WWW/K-12/airplane/mach.html
- http://www.astronautix.com/g/germanrocketplanes.html
- http://www.astronautix.com/w/winged.html
- http://www.luft46.com
- Air & Space Magazine, August/September 1987 "Eugen Sanger's Atmosphere Skipper of the 1940s Inspired the Space Plane," Smithsonian Institute,1987, ASIN-B000LCO4CW.
- Vectors website - http://vc.airvectors.net/idx_sci.html
- Russian WW-II rocket aircraft Bereznyak-Isayev BI-1, avail: http://www.gutenberg.us/articles/eng/Bereznyak-Isayev_BI-1
- NASA X-38, avail: https://www.nasa.gov/centers/dryden/multimedia/imagegallery/X-38/X-38_proj_desc.html
- http://bristolspaceplanes.com/library/your-spaceflight-manual/
- http://www.aerospaceweb.org/design/waverider/main.shtml
- https://web.archive.org/web/20061117015637/http://www.xcor.com/products/vehicles/ez-rocket.html
- https://www.faa.gov/about/office_org/headquarters_offices/ast/
- Energiya-Buran: The Soviet Space Shuttle.

Chichester, UK: Praxis Publishing Ltd. pp. 379–381. ISBN 978-0-387-69848-9.
- Astronauts fly with SpaceX in landmark launch for commercial spaceflight". Spaceflight Now. 16 November 2020.
- Boeing Starliner test flight planned for spring 2022".SpaceNews. 2021-12-20.
- "The Shape of Things to Come – Orbital's Prometheus™ Space Plane Ready for NASA's Commercial Crew Development Initiative. https://www.northropgrumman.com/#search=prometheus
- NASA'S MANAGEMENT OF CREW TRANSPORTATION TO THE INTERNATIONAL SPACE STATION
- Wikipedia, various.
- http://www.astronautix.com/g/germanrocketplanes.html

## Glossary of terms

ABMA – Army Ballistic Missile Agency, Redstone Arsenal, Huntsville, Alabama.
AFB – Air Force base.
AGC – Apollo Guidance Computer
AIAA – American Institute of Aeronautics and Astronautics.
ALU – arithmetic logic unit
AMD – Aircraft Missiles Division, Fairchild Hiller, Hagerstown, MD.
Antipodal – point on the Earth diametrically opposite.
AOMC – Army Ordnance Missile Command – 1958, Redstone Arsenal
Apogee – farthest point in the orbit from the Earth.
APU – auxiliary power unit
ARPA – Advanced Research projects Agency.
ASC – Advanced Spacecraft Computer, by IBM, for Titan launch vehicle.
ASIN – Amazon Standard Inventory Number
AST – FAA Office of Commercial Space Transportation
Astrionics – electronics for space flight.
ATOLL - Acceptance Test or Launch Language
Avatar - Aerobic Vehicle for Transatmospheric Hypersonic Aerospace Transportation (India)
BEO – beyond Earth orbit.
BFS – Shuttle backup flight software, see also pass
Blooster – a high altitude balloon as a first stage to orbit.
BP – boilerplate. Mechanical model.
Buran – Russian space shuttle.
CASC - China Aerospace Science and Technology Corp.

C3PO – (NASA) Commercial Crew and Cargo Payload Office.
CciCap – NASA's Commercial Crew Integrated Capability program.
COTS – commercial off-the-shelf; Commercial orbital transportation service.
Cpu – central processing unit.
CRS – Commercial Resupply Services (for ISS).
CRV – crew return vehicle
C-stoff – German rocket fuel, 30% hydrazine, 57% methanol, 13% water.
CTOL – conventional take off and landing.
CTS – Crew Transportation System
Cyrogenic – pertaining to very low temperatures.
DARPA – (U.S.) Defense Advanced Research Projects Agency.
Dead-stick – no controls. Refers to the control stick in an aircraft.
DoD – Department of Defense.
DLR – Deutsches Zentrum für Luft- und Raumfahrt, German Space Agency.
DPS – deorbit propulsion system
DTM – dynamic test model, for structural tests.
EADS – European Aeronautic Defense and Space Company, Airbus Industries
EAF – extra atmospheric flight
ECLSS – Environmental Control and Life Support system.
EELV – Evolved Expendable Launch Vehicle, Delta-IV and Atlas-V.
Ephemeris – position information data set for orbiting

bodies, 6 parameters plus time.
EPOS - (Russian) Experimental Passenger Orbital aircraft.
ESA – European Space Agency
ETR – U. S. Eastern Test Range, Kennedy launch site.
EVA – extra-vehicular activity – going outside in a space suit.
FAA – (U. S.) Federal Aviation Administration.
FAR – (U.S.) Federal Acquisition Regulations.
GAO – U. S. Government Accountability Office.
Gimbal – pivoted support, allowing rotation about 1 axis.
GPC – Shuttle general purpose computer, AP-101s
Gpm – gallons per minute.
GSFC – NASA Goddard Space Flight Center, Greenbelt, MD.
Gyro – device to measure angular rate.
Hermes – European Space plan project; canceled.
HOTOL – Horizontal take-off and landing (like an airplane)
HP – horsepower, 745.7 watts. From James Watt, to relate steam engines to the horses they replaced.
HTHL – horizontal take-off and landing.
Hypergolic – self-igniting.
Hypersonic – Mach 5 or above
ICBM – Intercontinental Ballistic Missile.
IBM – International Business Machines Company.
IRBM – Intermediate Range Ballistic Missile.
ISBN – international standard book number.
ISP – specific impulse. Measure of efficiency of rocket engine. Units of seconds.
ISRO – Indian Space Research Organization

IU –Saturn Instrument Unit.
JATO 0 jet assisted take off; actually using a sold fueled rocket engine.
JPL – Jet Propulsion Laboratory, Pasadena, CA
JSC – Johnson Space Center, Houston, Texas.
Karman line – 100 km, the "official" beginning of space.
KN – kilo newton, unit of force.
KPH – kilometers per hour
KSC – NASA Kennedy Space Center, launch site, Florida.
L2 – second of 5 Lagrange points, a null in the gravity field in the restricted 3-body problem.
Lbf – pounds, force.
LC-39 – Launch Complex – 39 at KSC.
LEM – Apollo lunar excursion module.
LEO – low Earth orbit.
LH2 – liquid hydrogen.
Lifting body – the entire underside of the aircraft is the wing.
LOX – liquid oxygen, boils at -297 F.
LVDA – Launch Vehicle Data Adapter.
LVDC – Launch Vehicle Digital Computer.
Mach number – speed of sound at the ambient temperature and pressure.
MBB – Messerschmidt-Bolekow-Bloehm, Germany aircraft company
Mev – million electron volts, measure of energy of a particle.
MINITRACK – "Minimum Trackable Satellite " U. S. satellite tracking network, 1957.
MIT – Massachusetts Institute of Technology.

MOL – USAF manned program. Canceled.
Mothership – carries another aircraft to altitude.
MPH – miles per hour.
MSC – Manned Space Center, Houston, TX. Renamed Johnson Space Center (JSC).
MSFC – NASA Marshall Space Flight Center, Huntsville, AL.
m/s – meters per second.
NACA – National Advisory Committee for Aeronautics, predecessor to NASA.
NASA – National Aeronautics and Space Administration.
NASCOM – NASA Communications Network. Worldwide, operated by GSFC.
NASP – National Aero-Space Plane
Newton – SI unit of force, 0.2248 lb-f.
NORAD – North American Air Defense.
NRL – Naval Research Lab, Washington, DC.
NRO – (U. S.) National Reconnaissance Office.
NTIS – National Technical Information Service (www.ntis.gov).
Ogive – a pointed arch shape.
OTV – orbital test vehicle
OV – orbiting vehicle.
PASS – (Shuttle) Primary Avionics System Software. See also BFS
Perigee – closest point in the orbit from the Earth.
PGNCS – Primary Guidance, Navigation, and Control System for Apollo.
POGO – longitudinal oscillation in liquid-fueled rocket motors that can lead to failure.
Pregnant Guppy – large cargo aircraft operated by Aero

Spacelines 1963-1979.
Pulsejet – jet engine that doesn't need to be moving forward to operate.
R&D – research & development.
RCS – reaction control system, rocket or cold gas thrusters for attitude control.
Redstone Arsenal – Army R&D facility in Huntsville, AL. Later became NASA MSFC.
RFNA – red fuming nitric acid
RP-1 – rocket propellant-one, highly refined kerosene.
RLV – reusable launch vehicle.
Sabre - Synergistic Air-Breathing Rocket Engine, a rocket-jet engine hybrid.
SAE – Society of Automotive Engineers (standards)
SAO – Smithsonian Astrophysical Observatory.
Scramjet – an air breathing jet engine designed for hypersonic flight.
Sanger-II – a proposed German rocket plane
SI – System International – the metric system.
S-IC – first stage of the Saturn V
S-II – second stage of the Saturn V
S-IVB – third stage of the Saturn V
S-IV – second stage of Saturn 1 rocket.
Silbervogel – (German) Silver Bird, Sanger's circa 1930 proposed rocket plane.
SLS – Space Launch System.
SMTS – Shuttle Mission Training Facility
SPACE – "Spurring Private Aerospace Competitiveness and Entrepreneurship" Act, by Congress, 2015.
SSTO – single stage to orbit.
STADAN – Space Tracking and Data Acquisition

Network.
STEM – Science, Technology, Engineering, Math curricula K through 12 level.
STS – U. S. Space Transportation system (Shuttle)
Titan – ICBM and NASA/USAF launch vehicle.
TM – Technical Manual.
Transonic – faster than the speed of sound at the ambient temperature and pressure.
TSO – two stage to orbit.
T-stoff – German oxidizer, high concentration hydrogen peroxide
UK – United Kingdom; England
Ullage – residual fuel or oxidizer in a tank after engine burn is complete.
V-2 – German World War-II ballistic missile developed by the von Braun Team.
VSS – Virgin Space Ship, built by Scaled Composites for Virgin Galactic.
VTHL – vertical take-off, horizontal landing.
VTVL – vertical takeoff, vertical landing.
VTO – vertical take-off.
Walc – water-alcohol mixture used as fuel.
WSMR – White Sands Missile Range, New Mexico.
WTR – U.S. Western Test Range, Vandenburg AFB in California launch site.
XCOR – American private aerospace company.
XOV – experimental orbital vehicle, part of the Blackstar project.
X-plane – experimental aircraft.

**If you enjoyed this book, you might be interested in some of his others as well.**

Stakem, Patrick H. *Floating Point Computation*, 2013, PRRB Publishing, ISBN-152021619X.

Stakem, Patrick H. *Architecture of Massively Parallel Microprocessor Systems*, 2011, PRRB Publishing, ISBN-1520250061.

Stakem, Patrick H. *Multicore Computer Architecture*, 2014, PRRB Publishing, ISBN-1520241372.

Stakem, Patrick H. *Personal Robots*, 2014, PRRB Publishing, ISBN-1520216254.

Stakem, Patrick H. *RISC Microprocessors, History and Overview*, 2013, PRRB Publishing, ISBN-1520216289.

Stakem, Patrick H. *Robots and Telerobots in Space Applications*, 2011, PRRB Publishing, ISBN-1520210361.

Stakem, Patrick H. *The Saturn Rocket and the Pegasus Missions, 1965*, 2013, PRRB Publishing, ISBN-1520209916.

Stakem, Patrick H. *Visiting the NASA Centers, and Locations of Historic Rockets & Spacecraft*, 2017, PRRB Publishing, ISBN-1549651205.

Stakem, Patrick H. *Microprocessors in Space*, 2011, PRRB Publishing, ISBN-1520216343.

Stakem, Patrick H. Computer *Virtualization and the Cloud*, 2013, PRRB Publishing, ISBN-152021636X.

Stakem, Patrick H. *What's the Worst That Could Happen? Bad Assumptions, Ignorance, Failures and Screw-ups in Engineering Projects, 2014,* PRRB Publishing, ISBN-1520207166.

Stakem, Patrick H. *Computer Architecture & Programming of the Intel x86 Family, 2013,* PRRB Publishing, ISBN-1520263724.

Stakem, Patrick H. *The Hardware and Software Architecture of the Transputer*, 2011,PRRB Publishing, ISBN-152020681X.

Stakem, Patrick H. *Mainframes, Computing on Big Iron*, 2015, PRRB Publishing, ISBN- 1520216459.

Stakem, Patrick H. *Spacecraft Control Centers*, 2015, PRRB Publishing, ISBN-1520200617.

Stakem, Patrick H. *Embedded in Space,* 2015, PRRB Publishing, ISBN-1520215916.

Stakem, Patrick H. *A Practitioner's Guide to RISC Microprocessor Architecture*, Wiley-Interscience, 1996,

ISBN-0471130184.

Stakem, Patrick H. *Cubesat Engineering*, PRRB Publishing, 2017, ISBN-1520754019.

Stakem, Patrick H. *Cubesat Operations*, PRRB Publishing, 2017, ISBN-152076717X.

Stakem, Patrick H. *Interplanetary Cubesats*, PRRB Publishing, 2017, ISBN-1520766173 .

*Stakem, Patrick H. Cubesat Constellations, Clusters, and Swarms, Stakem,* PRRB Publishing, 2017, ISBN-1520767544.

Stakem, Patrick H. *Graphics Processing Units, an overview*, 2017, PRRB Publishing, ISBN-1520879695.

Stakem, Patrick H. *Intel Embedded and the Arduino-101, 2017,* PRRB Publishing, ISBN-1520879296.

Stakem, Patrick H. *Orbital Debris, the problem and the mitigation*, 2018, PRRB Publishing, ISBN-*1980466483*.

Stakem, Patrick H. *Manufacturing in Space*, 2018, PRRB Publishing, ISBN-1977076041.

Stakem, Patrick H. *NASA's Ships and Planes*, 2018, PRRB Publishing, ISBN-1977076823.

Stakem, Patrick H. *Space Tourism*, 2018, PRRB

Publishing, ISBN-1977073506.

Stakem, Patrick H. *STEM – Data Storage and Communications*, 2018, PRRB Publishing, ISBN-1977073115.

Stakem, Patrick H. *In-Space Robotic Repair and Servicing*, 2018, PRRB Publishing, ISBN-1980478236.

Stakem, Patrick H. *Introducing Weather in the pre-K to 12 Curricula, A Resource Guide for Educators*, 2017, PRRB Publishing, ISBN-1980638241.

Stakem, Patrick H. *Introducing Astronomy in the pre-K to 12 Curricula, A Resource Guide for Educators*, 2017, PRRB Publishing, ISBN-198104065X.
Also available in a Brazilian Portuguese edition, ISBN-1983106127.

Stakem, Patrick H. *Deep Space Gateways, the Moon and Beyond*, 2017, PRRB Publishing, ISBN-1973465701.

Stakem, Patrick H. *Exploration of the Gas Giants, Space Missions to Jupiter, Saturn, Uranus, and Neptune*, PRRB Publishing, 2018, ISBN-9781717814500.

Stakem, Patrick H. *Crewed Spacecraft*, 2017, PRRB Publishing, ISBN-1549992406.

Stakem, Patrick H. *Rocketplanes to Space*, 2017, PRRB Publishing, ISBN-1549992589.

Stakem, Patrick H. *Crewed Space Stations,* 2017, PRRB Publishing, ISBN-1549992228.

Stakem, Patrick H. *Enviro-bots for STEM: Using Robotics in the pre-K to 12 Curricula, A Resource Guide for Educators,* 2017, PRRB Publishing, ISBN-1549656619.

Stakem, Patrick H. *STEM-Sat, Using Cubesats in the pre-K to 12 Curricula, A Resource Guide for Educators*, 2017, ISBN-1549656376.

Stakem, Patrick H. *Lunar Orbital Platform-Gateway*, 2018, PRRB Publishing, ISBN-1980498628.

Stakem, Patrick H. *Embedded GPU's*, 2018, PRRB Publishing, ISBN- 1980476497.

Stakem, Patrick H. *Mobile Cloud Robotics*, 2018, PRRB Publishing, ISBN- 1980488088.

Stakem, Patrick H. *Extreme Environment Embedded Systems,* 2017, PRRB Publishing, ISBN-1520215967.

Stakem, Patrick H. *What's the Worst, Volume-2*, 2018, ISBN-1981005579.

Stakem, Patrick H., *Spaceports*, 2018, ISBN-1981022287.

Stakem, Patrick H., *Space Launch Vehicles*, 2018, ISBN-1983071773.

Stakem, Patrick H. *Mars*, 2018, ISBN-1983116902.

Stakem, Patrick H. *X-86, 40$^{th}$ Anniversary ed*, 2018, ISBN-1983189405.

Stakem, Patrick H. *Lunar Orbital Platform-Gateway*, 2018, PRRB Publishing, ISBN-1980498628.

Stakem, Patrick H. *Space Weather*, 2018, ISBN-1723904023.

Stakem, Patrick H. *STEM-Engineering Process*, 2017, ISBN-1983196517.

Stakem, Patrick H. *Space Telescopes,* 2018, PRRB Publishing, ISBN-1728728568.

Stakem, Patrick H. *Exoplanets*, 2018, PRRB Publishing, ISBN-9781731385055.

Stakem, Patrick H. *Planetary Defense*, 2018, PRRB Publishing, ISBN-9781731001207.

Patrick H. Stakem *Exploration of the Asteroid Belt*, 2018, PRRB Publishing, ISBN-1731049846.

Patrick H. Stakem *Terraforming*, 2018, PRRB Publishing, ISBN-1790308100.

Patrick H. Stakem, *Martian Railroad,* 2019, PRRB Publishing, ISBN-1794488243.

Patrick H. Stakem, *Exoplanets,* 2019, PRRB Publishing, ISBN-1731385056.

Patrick H. Stakem, *Exploiting the Moon,* 2019, PRRB Publishing, ISBN-1091057850.

Patrick H. Stakem, *RISC-V, an Open Source Solution for Space Flight Computers,* 2019, PRRB Publishing, ISBN-1796434388.

Patrick H. Stakem, *Arm in Space*, 2019, PRRB Publishing, ISBN-9781099789137.

Patrick H. Stakem, *Extraterrestrial Life*, 2019, PRRB Publishing, ISBN-978-1072072188.

Patrick H. Stakem, *Space Command*, 2019, PRRB Publishing, ISBN-978-1693005398.

CubeRovers, A Synergy of Technologys, 2020, PRRB Publishing, ISBN-979-8651773138.

Robotic Exploration of the Icy moons of the Gas Giants. 2020, PRRB Publishing, ISBN- 979-8621431006

Hacking Cubesats, 2020, PRRB Publishing, ISBN-979-8623458964.

History & Future of Cubesats, PRRB Publishing, ISBN-979-8649179386.

Hacking Cubesats, Cybersecurity in Space, 2020, PRRB Publishing, ISBN-979-8623458964.

Powerships, Powerbarges, Floating Wind Farms: electricity when and where you need it, 2021, PRRB Publishing, ISBN-979-8716199477.

Hospital Ships, Trains, and Aircraft, 2020, PRRB Publishing, ISBN-979-8642944349.

*CubeRovers, a Synergy of Technologys*, 2020, ISBN-979-8651773138

*Exploration of Lunar & Martian Lava Tubes by Cube-X*, ISBN-979-8621435325.

*Robotic Exploration of the Icy moons of the Gas Giants*, ISBN- 979-8621431006.

*History & Future of Cubesats*, ISBN-978-1986536356.

*Robotic Exploration of the Icy Moons of the Ice Giants, by Swarms of Cubesats*, ISBN-979-8621431006.

*Swarm Robotics,* ISBN-979-8534505948.

*Introduction to Electric Power Systems*, ISBN-979-

8519208727.

*Centros de Control: Operaciones en Satélites del Estándar CubeSat* (Spanish Edition), 2021, ISBN-979-8510113068.

*Exploration of Venus*, 2022, ISBN-979-8484416110.

Patrick H. Stakem, *The Search for Extraterrestial Life,* 2019, PRRB Publishing, ISBN-1072072181.

*The Artemis Missions, Return to the Moon, and on to Mars,* 2021, ISBN-979-8490532361.

*James Webb Space Telescope. A New Era in Astronomy*, 2021, ISBN-979-8773857969.

www.ingramcontent.com/pod-product-compliance
Lightning Source LLC
Chambersburg PA
CBHW020929180526
45163CB00007B/2946